I0813574

HACKERS: BEHIND THE CODE

12 INCREDIBLE FACTS ABOUT CYBER-SECURITY

BLACK RABBIT BOOKS

Table of Contents

Cybersecurity Protects from *Hackers*

1

Criminals rob banks and stores. But did you know some thieves steal online? These people are known as hackers. They steal valuable data. Some run **scams**. They target people. They want access to online accounts. If the scam is successful, they control those accounts.

Hackers can also focus on companies. They use computer skills to enter computer systems. This allows them to steal many private records at once. Sometimes they sell the stolen data. Or they demand money to recover it. In some cases, hackers threaten to erase the data. Once data is erased, a company might not get it back. This can affect their business and customers.

How do people and companies protect themselves? They use cybersecurity. It protects devices and data. It prevents hackers from getting to sensitive information.

Companies and people can share in this important work. Strong passwords are part of cybersecurity. Workers in this field are strong defenders of data. They create secure online spaces. This field grows more important each day. More and more devices rely on the internet. The ways that people can hack private data grows as well.

HACKER
ATTACK

2 Anything with Internet *Can Be Hacked*

Think about your daily life. How many things use the internet? Some things are clear. Tablets and computers use the internet. Most phones do as well. But what about your TV? Your watch? The lights in your house? The doorbell? The internet has made many things easier to use. But that ease comes with a price: the ability to hack into them.

Why would someone want to hack a doorbell? One reason is to use the camera in the doorbell. If they can view your front porch, they can watch your home. They know what you and your family members look like. They know when you leave and come back. They can steal private information and plan illegal activity.

Some hackers look for bigger challenges. It is surprisingly simple to change highway signs. Hackers might change a sign as a prank. However, there are

Think About It Make a list of everything in your house that uses the internet. If a hacker got into these devices, what could they do?

Hacking home devices is like stealing someone's keys to their home.

more serious targets. Hackers have hacked oil pipelines. This stopped the flow of fuel supplies. Hackers even pose a big threat to the military. The F-35 is known as "the flying computer." A hacker could change its flight path or crash the jet. They could take over the weapons system. They can create confusion and damage in many ways.

The F-35 is a supersonic stealth strike fighter jet.

2 in 3 American gamers who have fallen for scams.

Hackers may use cheat codes to get into your system. • They steal credit card data and private information. • On average, thieves steal $330 through video game hacks.

Cybersecurity Keeps *Data Safe*

3

Think about how many devices are in your home. Now, think about how many houses are on your street. Each of those has things using the internet. Multiply that by the houses in your town. Or the whole country! That's a lot of devices to protect.

Most businesses use the internet as well. Even places you might not think of, like the dentist. Think of your last visit. What happened first? Someone checked you in on a computer. Did you need a cleaning? An assistant looked up your dental records. They made new notes about your teeth. Any past X-rays were **synced** with your file. All these processes use the internet. Someone needs to keep these records safe. You don't want just anyone to have access.

A hack can majorly affect the flow of information. And it affects more than tech companies. In 2021, a meat-packing company was the target. It shut down beef and poultry plants on four continents. They paid $11 million to settle the attack!

Cybersecurity lets people use the internet safely. It keeps data safe. This lets workers make everyday tasks as easy as possible. Plus, they keep our gaming, apps, and everything else we do online secure and worry-free!

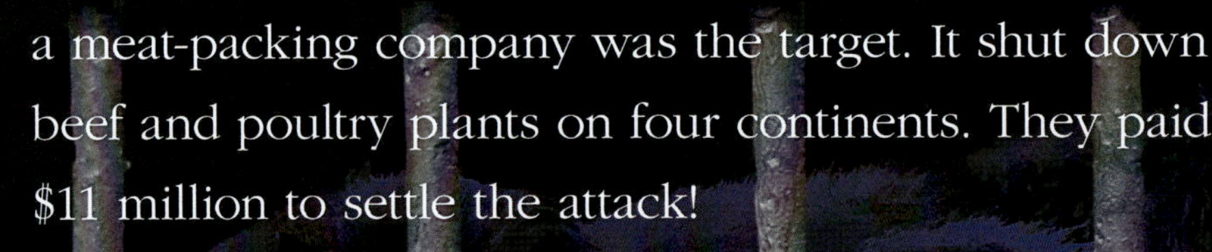

$10.5 trillion How much experts think cybercrime will cost in 2025.

This number is more than just stolen funds. It includes time spent recovering data and fixing issues caused by hackers. • Criminals make more money through online crime than any other kind of crime. • Companies must spend time and resources to repair the damage caused by a **breach**.

Even farms and agriculture can be the target of a hacker attack.

Think About It Imagine the internet disappeared. How would it affect your daily life?

Small Businesses and Elderly People *Are More at Risk*

4

Many movies and TV shows feature hackers. Sometimes, they show hackers interacting with someone. Then they hack into that person's devices. This implies that hackers target specific people. Hacking someone looks like it is personal or a form of revenge.

This is not always true. Some hackers do target specific people. But most are not a personal attack. Hackers don't have time. They are looking for **profiles**. This is a set of shared qualities. Large groups of people typically fit a profile.

The biggest targets are small businesses and older people. These groups have less secure online habits. Small businesses struggle with tight budgets. They do not have a lot of money to spend on cybersecurity. Older people are not always aware of cybercrime.

Older people are more likely to trust scammers and share sensitive information.

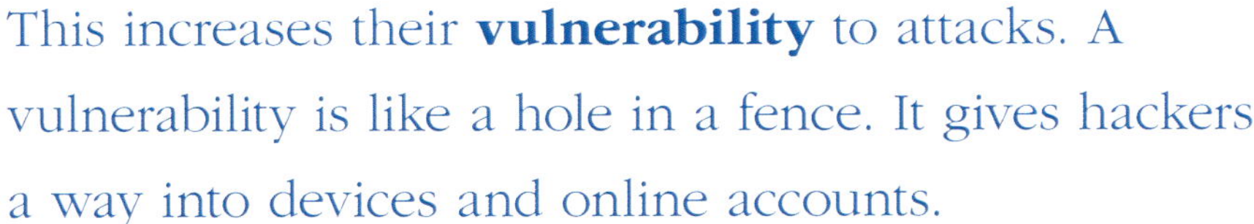

This increases their **vulnerability** to attacks. A vulnerability is like a hole in a fence. It gives hackers a way into devices and online accounts.

Hackers often use email to create a vulnerability. Ninety-one percent of attacks begin with an email. These are called **phishing** attacks. At first, a hacker creates an email that looks important. It may look like it is from a bank or IT help desk. People click on a link or open an attachment. This runs a **script** that places **malware** on the device. That's all it takes to give a hacker access.

39 Average time in seconds between hacker attacks worldwide.

Some scripts use common login information to try to break into a computer. • These are known as "brute force attacks." • Once past the login, hackers can use the computer to steal data.

Cybersecurity Must Adapt *Quickly*

5

Hackers always create new ways to break into systems. They improve their methods. They update coding scripts. They also design new emails to trick people. For cybersecurity workers, it is hard to stay on top of the latest hacking trends. There is a common saying in the field. "It's not a matter of if you're going to get hacked; it's when." Cybersecurity does not know when an attack will happen. But they are always prepared.

Many FBI Special Agents have a background in technology, including cybersecurity.

The best way to prepare is to create layers of defense. Think of it like you're putting on clothes for a cold day. First, you put on a shirt. This is your base layer. Next you add a sweater. Then you put on your coat. This is the thickest layer. It blocks most of the cold. But the other layers are just as important. You wouldn't put on one layer

and expect to stay warm. They all prevent your body from getting cold.

That is how cybersecurity works too. It builds layers of defense. A hacker might be able to break through one layer. Maybe they get through two layers. But they will likely give up if they run into a third or fourth layer. Remember, hackers prefer easy targets. They look for vulnerabilities.

"IT'S NOT A MATTER OF IF YOU'RE GOING TO GET HACKED; IT'S WHEN."

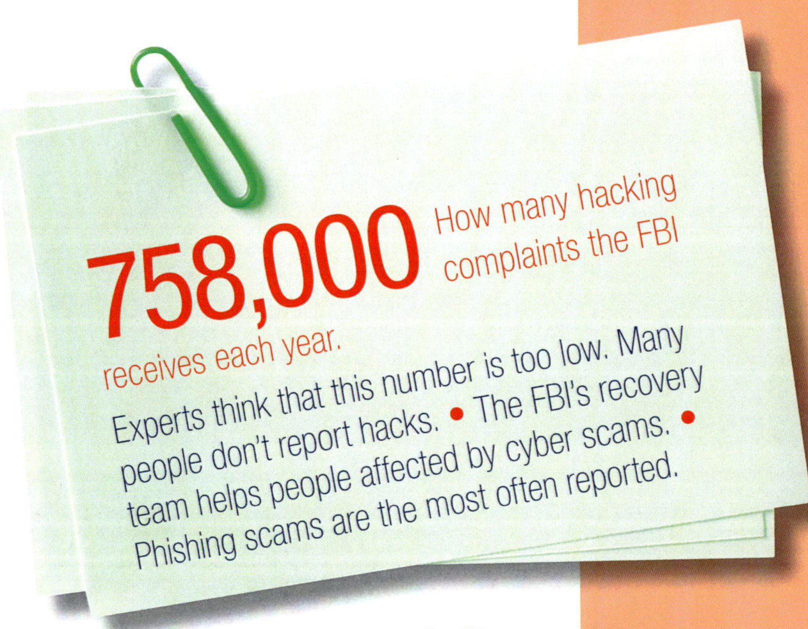

Layers are the key for the best protection.

LET'S (NOT) GO PHISHING

Phishing scams pretend to be a real company. They come from texts, emails, or phone calls. They look real at first. But there will be signs that something is wrong. Look for words that are spelled wrong. Sentences may sound awkward. The email address may not match who they claim to be. Don't click on any attachments they send. Just report the message as SPAM.

Artificial Intelligence Is a Bodyguard *for Data*

6

As **artificial intelligence** (AI) becomes more common, it is used in more places. AI can be used to write papers. It can create images. It is even used in cybersecurity.

When a company is hacked, some hints appear in the company's networks. Those hints aren't always clear to humans. But AI can analyze huge amounts of data quickly. It pulls data from all available sources. It can quickly spot if something seems wrong. This kind of monitoring is called threat detection or intelligence.

46 Percent of data breaches involving personal identity information (PII).

PII is the most common kind of data stolen. • It includes information like tax ID numbers, home addresses, and more. • Hackers use PII for identity theft and credit card theft.

AI can be used in other ways. It can detect an

anomaly. AI monitors activity on a company's networks. It learns what normal activity looks like. When something big changes, AI flags it. Then a human checks for a threat. AI can also examine user behavior. It learns where people log in and what data they access. It flags if someone logs in from a strange place or accesses sensitive data. Plus, AI can respond to threats without help from a human. This technology helps shut down systems so hackers can't get in. Then it sends alerts to human team members. This gives a recovery team valuable time to respond.

AI can be like a copilot for cybersecurity. It can help them a lot, but it is not perfect. Human judgment is still necessary. AI can send false alerts or false positives. AI processing is only as good as the data it uses.

PROTECTOR AND THREAT? It is not only the good guys who use AI. Hackers also use AI. They create attacks and scams using **deepfakes**. Deepfakes make it even harder to avoid phishing scams. How can you tell if something (or someone) is real? In a video, look closely at the mouth. See if the movements match the audio. In photos, look for differences in skin tone. Many deepfakes just swap heads with bodies.

The Safest Place for Data *Is Underground*

7

Data is made up of facts and details. There is private and public data. A lot of this information is stored on computers and networks. It is very valuable to hackers. They can make a lot of money and do much damage if they gain access to certain data. But where is this data?

The answer varies. On a cell phone, data is stored in files on the device. But the data might also sync to a server, like Apple's iCloud or Google Drive. So where do Apple and Google store that data? The answer is a data center. Some people call it "the **cloud**." Cloud data is often stored in underground data centers.

It might seem strange at first. But keeping these large servers underground makes a lot of sense. Large servers give off a lot of heat. It is cooler underground. The location also keeps the data safe from natural

Servers store huge amounts of data, including customer information, financial records, and more.

disasters, like tornados and hurricanes. They are even protected from earthquakes. Most spaces are at least 100 feet (30 m) below ground. There is less shaking this far down. The structures also use surrounding rocks as natural shields. These rocks help absorb the quaking.

These data centers have few entrances. They are safe from intruders and air strikes. If the goal is to back up data, it means that a copy is always there. This is a big relief for cybersecurity workers.

Think About It

Google is exploring the idea of building data centers underwater! What do you think are the pros and cons of this?

Iron Mountain is in an old limestone mine.

Some Workers Are *Former Hackers*

8

"Hackers" does not always refer to criminals. There are good and bad hackers. Some people's job is to hack into secure systems.

"Black hat" hackers are criminals. They steal data and try to hack into systems. On the other side, there are "white hat" hackers. They try to stop the bad hackers. These hackers are very similar in their actions. They both code scripts to break into systems. They look for ways to steal data. But white hat hackers do it for **ethical** reasons. Black hat hackers do not.

Companies hire white hat hackers to test cybersecurity. These hackers have job titles. They are called ethical hackers or cyber ninjas. Those who work for the government may be called special agents. These ethical hackers are paid to find vulnerabilities.

They play hide-and-seek, looking for unethical hackers in computer systems.

Many white hat hackers have black hat pasts. Kevin Mitnick was the world's first cybercriminal. He broke into the most secure networks in the world. He was eventually caught by the FBI and sent to prison. After his release, he worked in cybersecurity.

Black hat and white hat hackers use similar skills while on opposite sides of the law.

1940 Year Rene Carmille became the first ethical hacker.

Carmille was a member of the French Resistance in World War II. • He was an expert in punch-card computers, one of the earliest computers. • He let Nazis use his computers to find Jews. Then he secretly hacked them and spoiled their efforts.

NOT JUST BLACK AND WHITE "Black hat" and "white hat" refer to Old West movies. In these movies, the good guys wore white hats. The bad guys wore black hats. The color is meant to describe their intentions—good or bad. There are also "gray hats." These hackers have good intentions. But they do not have permission to hack into a system.

Not All Roles Call for *Coding Skills*

9

Jobs in cybersecurity need specific skills. However, not all workers need to know how to code or write scripts. Thinking like a hacker is the most important part.

One tool in a hacker's toolkit is **social engineering**. This does not even use a computer. It involves persuading people to share their secrets. A hacker may test an employee. They pretend to be an IT support person. They call the employee and tell a fake story. Then they ask the employee to share their login information. With that information, the hacker can enter the company's network.

All the coding skills in the world won't stop a social engineering attack. This is why some roles focus on non-technical skills. Workers learn critical thinking and problem-solving. Cybersecurity roles must often think creatively. They need to predict a hacker's next move.

These roles also need communication skills. It is important to be able to share information about attacks. This helps people understand how to avoid future attacks. Many tech skills can be learned on the job. Cybersecurity is about learning quickly and adapting.

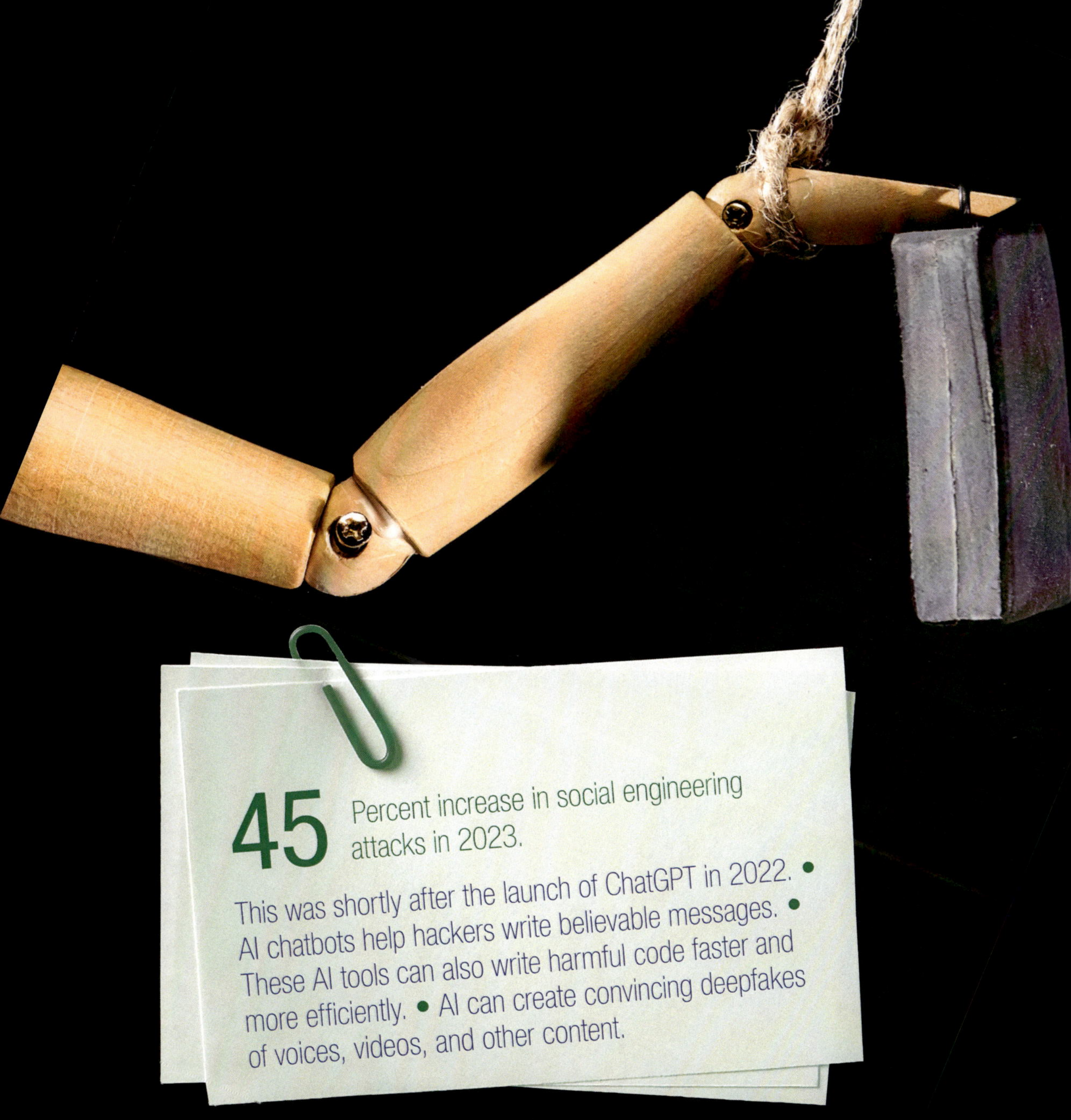
45 Percent increase in social engineering attacks in 2023.
This was shortly after the launch of ChatGPT in 2022. • AI chatbots help hackers write believable messages. • These AI tools can also write harmful code faster and more efficiently. • AI can create convincing deepfakes of voices, videos, and other content.

Cybersecurity Requires Constant *Monitoring*

10

Hackers constantly change their scams. Cybersecurity must stay one step ahead. Data must always be monitored. Failure to do so can impact the company, workers, and customers.

Companies create plans for defense. Most cyber-defense plans use the 5 C's. These are pillars to create the plans. They include **c**hange, **c**ontinuity, **c**ost, **c**ompliance, and **c**overage. *Change* is related to time. Cybersecurity is always changing. This pillar addresses the need to adapt. Workers must stay ahead of threats. *Continuity* means something that continues. If an attack strikes, can the business continue? What areas will be affected? This pillar creates a backup. It makes a recovery plan.

Cybersecurity can be costly. But an attack can cost even more. That's why the third C is *cost*. Spending money keeps data safe. How much can the business afford

change

continuity

cost

compliance

coverage

Think About It

Read through the 5 C's again. Is there a sixth C that you would add to the plan?

292 Number of days it takes to discover and fix a data breach.

The longer a breach goes unnoticed, the more money it takes to fix. • A breach lasting over 200 days can cost more than $5 million. • The most common causes are hacker attacks, IT failure, and human error.

to spend? The next C is *compliance*. Companies must follow certain laws. These laws keep private data safe. This pillar is a promise to obey the laws. Workers check that all requirements are met. The last C is *coverage*. How does the company protect itself? This pillar outlines how the company secures networks and data.

Strong Passwords Stop *Hackers*

11

Keeping secure online might sound impossible. But there is one easy thing you can do. And it is one of the best ways to stop hackers. How? Choose a strong password.

Hackers guess common passwords quickly. They have scripts that try hundreds a minute. And these scripts pull from resources that collect data about you. A hacker can also write a script. To do this, they find your first and last name. They may find your birthday on social media. Then they may look up your address. They can find all this information online.

Then the hacker will try any combination of this data. Common usernames include names and locations. Common passwords might include your birthday or birth year. For example, Mary Smith was born in 2010. Her username may be msmith2010. Does she own a pet? A hacker can find out on social media. She owns a dog named Callie. Her password might have this name. It is very easy to guess.

Think about this next time you create a password. Try using different characters. Use random letters, numbers, and symbols. Remembering might be challenging. But you are far less likely to get hacked!

Always include letters, numbers, and symbols when creating passwords without using a password manager, and always incorporate a mix of upper- and lowercase letters.

People Are the Most *Important Defense*

12

The only thing we can control is our actions. That is why YOU are the best person to stop someone from hacking you. You can take smart steps for protection.

First, keep private information private. Think of a password like underwear. Would you share your underwear with anyone? No! Your passwords should be just as private.

Never reply to messages from people you do not know. A "wrong number" can be a scammer. Avoid public Wi-Fi networks. These are not as secure. Hackers may be watching and can steal anything you type.

Most importantly, create long, unique passwords. Make it yours (and impossible for anyone to guess). For extra protection, use **two-factor authentication**. This makes your login extra secure. Every smart step you take makes hacking much harder. Hackers look for weak spots. Every safe action helps keep them out.

Cyber threats change all the time, and so do the ways to stop them. Cybersecurity is about protecting data and stopping hackers. Experts work tirelessly to secure networks. They safeguard personal information. And they prevent cyber threats from spreading. They create a safer internet for everyone.

An unknown caller could be a scammer.

Some sites send you a code to make sure it's your account.

64 Percent of Americans who do not know how to protect themselves from being hacked.
Did you get hacked? Tell a trusted adult immediately. • Keep accounts with payment information extra secure. • Change your password anywhere you might have used that same password.

Fact

• Cyber forensics track down hackers. These experts use advanced technology to track, analyze, and recover digital evidence. They trace back through the hacker's steps. They comb through data, emails, IP addresses, and more. They look for anything that could identify the hacker.

• Not many US hackers are found and charged. Only around 0.05% face legal charges for their actions. Cybercriminals don't need to be in the same city, state or even country as the scene of their crimes. The process is complicated. And it can take a long time to track them down.

Sheet

- Ninety-seven percent of companies have seen an increase in cyber threats since the start of the Russia-Ukraine war in 2022. Stress between countries can upset industries. Hackers can take advantage of this chaos. They use it as a cover for illegal activity online.

- Most businesses, big or small, have an online presence. Many companies have switched to online sales. This makes cybersecurity extra important in today's market. Experts predict worldwide security spending will hit $314 billion by 2028.

Glossary

anomaly
Something that is unusual or unexpected.

artificial intelligence
An area of computer science that deals with giving machines the ability to seem like they have human intelligence.

breach
A gap in a barrier or defense where someone can attack.

cloud
Large computers you can connect to on the internet and use for storing data.

deepfake
An image, video, or voice recording edited or created by artificial intelligence to look like a real person or event.

ethical
Following accepted rules of behavior.

malware
Software that is designed to mess up, damage, or gain unauthorized access to a computer system.

phishing
The unlawful practice of sending messages that look believable in order to trick targets into giving out personal information.

profile
A digital representation of a person on the internet.

scam
A dishonest way to make money by lying to people.

script
A sequence of instructions or commands written in a programming language that is carried out by another program rather than the user.

social engineering
The use of lies and psychological influence to get someone to share secret or private information that may be used for unlawful purposes.

sync
The process of linking data on two or more devices to make sure it is consistent and up to date.

two-factor authentication
A security method where users must provide two different identification factors to get into an account or system.

vulnerability
A flaw or weakness in a system or process that can be used by an attacker.

For More Information

Books

Doeden, Matt. *Cybersecurity: Protecting Your Personal Information.* Minneapolis: Lerner Publications, 2025.

Henzel, Cynthia Kennedy. *Be a Cybersecurity Specialist.* San Diego, CA: BrightPoint Press, 2025.

London, Martha. *Cybersecurity.* Minneapolis: Bearport Publishing Company, 2023.

Websites

Cyber Safety Video Series
cyber.org/cybersafety

Cybersecurity 101
tpt.pbslearningmedia.org/resource/nvcy-sci-cyber101/cybersecurity-101/

Cybersecurity Lab
www.pbs.org/wgbh/nova/labs/lab/cyber/

About the Author

Meghan Hatalla is a Minnesota-based writer and UX researcher exploring the intersection of people, technology, and storytelling. She critically examines everything from enterprise AI to historical disasters to create engaging nonfiction for curious minds. When not writing, she's exploring forests, lifting heavy things, or chasing down obscure facts.

Index

TOP RANK is published by Black Rabbit Books, P.O. Box 227, Mankato, MN, 56002. • Copyright © 2026 Black Rabbit Books. All rights reserved. No part of this book may be reproduced in any form without written permission from the publisher. • Designed by Danny Nanos • Photographs © Alamy Stock Photo/ vincent abbey, 19; Dreamstime/Anna Chaplygina, 11, Phichitpon Intamoon, 4–5; Getty Images/Stephanie Strasburg/Bloomberg, 26; Shutterstock/Andrey_Popov, 21, a2iStokker, 41, Belinda Pretorius, 48, bookzv, 14, creativerse, 17, denisik11, 24–25, Denphumi, 36, Dmitry Demidovich, 22, Elnur, 37, 38, ene, 9, Gelpi, 39, HAKINMHAN, 31, Halfpoint, 15, hilalabdullah, 44, IdeaGeneration, 30, Jolygon, 46–47, kai keisuke, 2–3, 28–29, Kirill Kurashov, 27, kung_tom, 22, LightField Studios, 32–33, Linaimages, 40, NicoElNino, 45, Oleksandr Lytvynenko, 12, Pasuwan, 23, PBXStudio, 44, Philip Steury Photography, 20, Prostock-studio, 6, 23, ranchorunner, 10, RossiAgung, 45, Runrun2, 41, Sajjad Farooq, cover, 1, SkillUp, 17, Sorapop Udomsri, 39, tete_escape, 42–43, Thomas Marx, 12–13, Tomas Ragina, 7, ultramansk, 12–13, VCoscaron, 8, wabeno, 16, worker, 2 • Printed in the United States of America.

Library of Congress Cataloging-in-Publication Data: Names: Hatalla, Meghan author | Title: 12 incredible facts about cybersecurity / by Meghan Hatalla. | Description: Mankato, MN: Top Rank is an imprint of Black Rabbit Books, [2026] | Series: Hackers: behind the code | Includes bibliographical references and index. | Audience: Ages 9–13 | Audience: Grades 4–6 | Identifiers: LCCN 2025021464 (print) | LCCN 2025021465 (ebook) | ISBN 9781645825210 library binding | ISBN 9781645825395 paperback | ISBN 9781645825579 ebook | Subjects: LCSH: Computer security—Juvenile literature | Classification: LCC QA76.9.A25 .H384 2026 (print) | LCC QA76.9.A25 (ebook) | DDC 005.8—dc23/eng/20250716 | LC record available at https://lccn.loc.gov/2025021464